SCIENCE CHUNKS

Year C Student Notebook

Science Chunks - Year C Notebook

First Edition 2023

ISBN: 978-1-953490-22-3

Email: support@elementalscience.com

Copyright Policy

Table of Contents

SCIENCE CHUNKS

Living Body Unit

Living or Nonliving

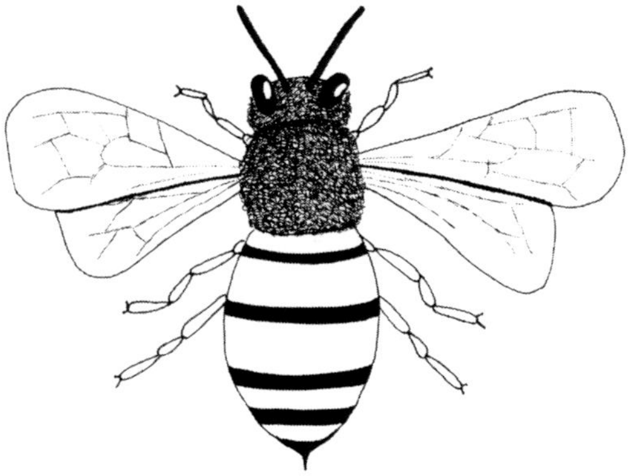

Single-celled Organisms

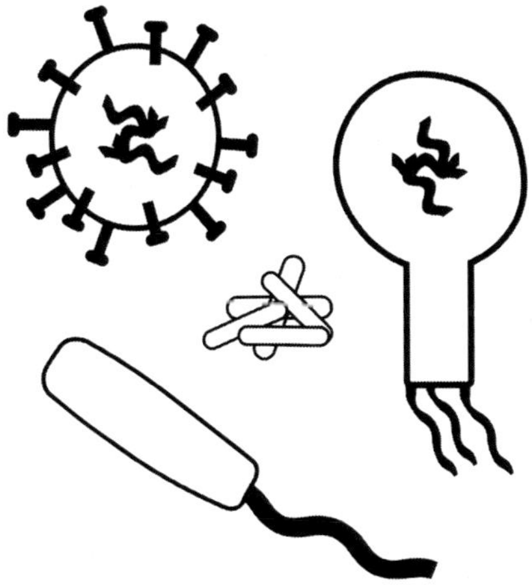

Fungi and Molds

Plants and Animals

SCIENCE CHUNKS

Invertebrates Unit

Mollusks

Crustaceans

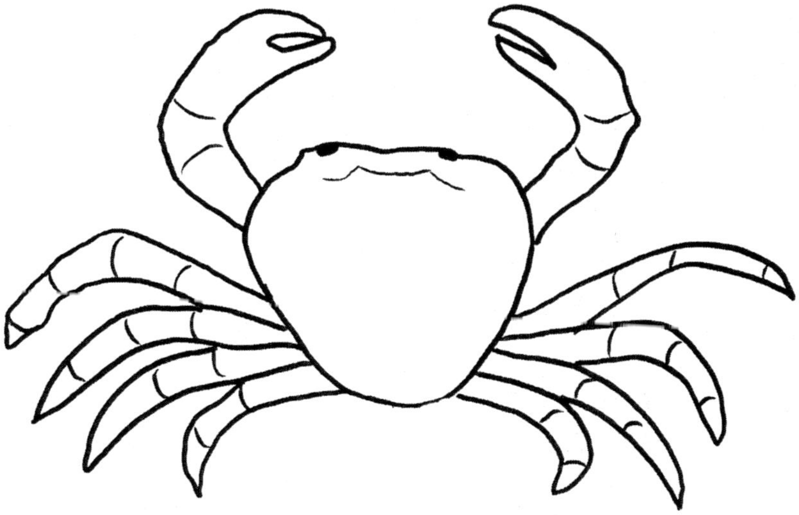

Arachnids and Cnidarians

Arachnids

Cnidarians

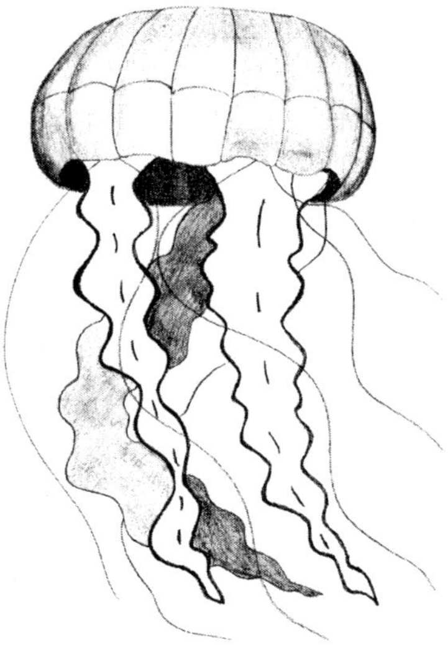

Invertebrate Life Cycles

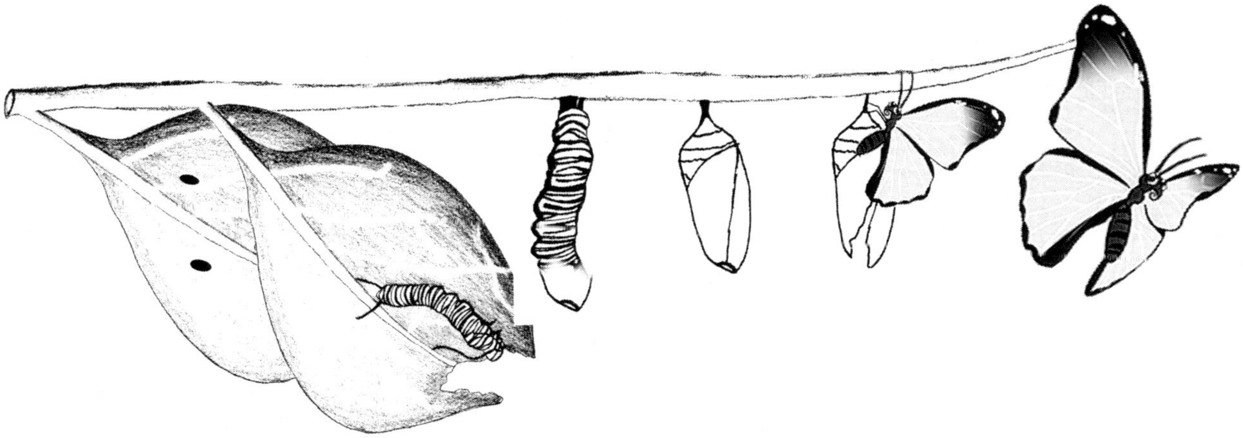

Insects

SCIENCE CHUNKS

Planet Earth Unit

Inside the Earth

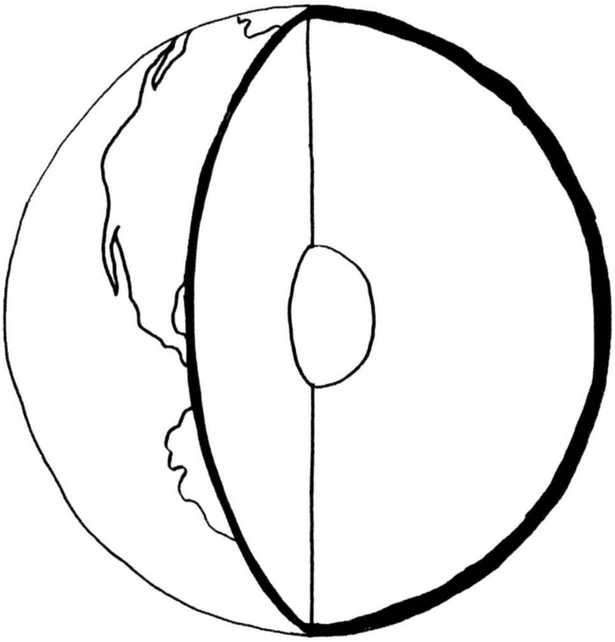

On the Earth

Mountains

Maps

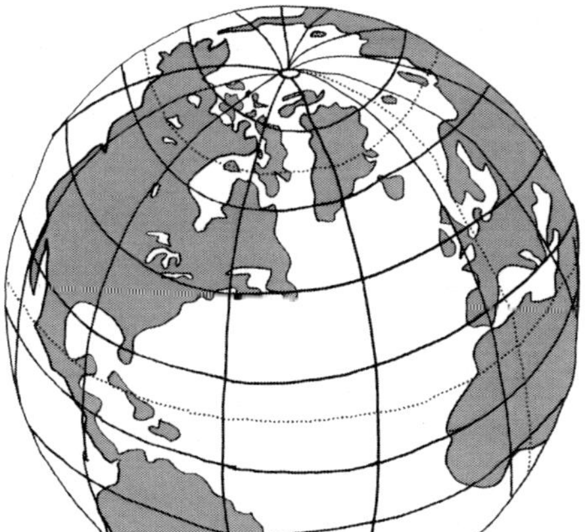

Conservation

SCIENCE CHUNKS

Periodic Table Unit

The Periodic Table

1 H Hydrogen 1.008																	2 He Helium 4.003
3 Li Lithium 6.941	4 Be Beryllium 9.012											5 B Boron 10.81	6 C Carbon 12.01	7 N Nitrogen 14.01	8 O Oxygen 16.00	9 F Fluorine 19.00	10 Ne Neon 20.18
11 Na Sodium 22.99	12 Mg Magnesium 24.31											13 Al Aluminum 26.98	14 Si Silicon 28.09	15 P Phosphorus 30.97	16 S Sulfur 32.07	17 Cl Chlorine 35.45	18 Ar Argon 39.95
19 K Potassium 39.10	20 Ca Calcium 40.08	21 Sc Scandium 44.96	22 Ti Titanium 47.87	23 V Vanadium 50.94	24 Cr Chromium 52.00	25 Mn Manganese 54.94	26 Fe Iron 55.85	27 Co Cobalt 58.93	28 Ni Nickel 58.69	29 Cu Copper 63.55	30 Zn Zinc 65.39	31 Ga Gallium 69.72	32 Ge Germanium 72.61	33 As Arsenic 74.92	34 Se Selenium 78.96	35 Br Bromine 79.90	36 Kr Krypton 83.80
37 Rb Rubidium 85.47	38 Sr Strontium 87.62	39 Y Yttrium 88.91	40 Zr Zirconium 91.22	41 Nb Niobium 92.91	42 Mo Molybdenum 95.94	43 Tc Technetium 98.91	44 Ru Ruthenium 101.1	45 Rh Rhodium 102.9	46 Pd Palladium 106.4	47 Ag Silver 107.9	48 Cd Cadmium 112.4	49 In Indium 114.8	50 Sn Tin 118.7	51 Sb Antimony 121.8	52 Te Tellurium 127.6	53 I Iodine 126.9	54 Xe Xenon 131.3
55 Cs Cesium 132.9	56 Ba Barium 137.3	* 71 Lu Lutetium 175.0	72 Hf Hafnium 178.5	73 Ta Tantalum 181.0	74 W Tungsten 183.9	75 Re Rhenium 186.2	76 Os Osmium 190.2	77 Ir Iridium 192.2	78 Pt Platinum 195.1	79 Au Gold 197.0	80 Hg Mercury 200.6	81 Tl Thallium 204.4	82 Pb Lead 207.2	83 Bi Bismuth 209.0	84 Po Polonium [209]	85 At Astatine [210]	86 Rn Radon [222]
87 Fr Francium [223]	88 Ra Radium [226]	** 103 Lr Lawrencium [262]	104 Rf Rutherfordium [261]	105 Db Dubnium [262]	106 Sg Seaborgium [266]	107 Bh Bohrium [264]	108 Hs Hassium [269]	109 Mt Meitnerium [268]	110 Ds Darmstadtium [272]	111 Rg Roentgenium [272]	112 Cn Copernicium [285]	113 Nh Nihonium [286]	114 Fl Flerovium [289]	115 Mc Moscovium [289]	116 Lv Livermorium [293]	117 Ts Tennessine [294]	118 Og Oganesson [294]

*Lanthanides

57 La Lanthanum 138.9	58 Ce Cerium 140.1	59 Pr Praseodymium 140.9	60 Nd Neodymium 144.2	61 Pm Promethium [145]	62 Sm Samarium 150.4	63 Eu Europium 152.0	64 Gd Gadolinium 157.3	65 Tb Terbium 158.9	66 Dy Dysprosium 162.5	67 Ho Holmium 164.9	68 Er Erbium 167.3	69 Tm Thulium 168.9	70 Yb Ytterbium 173.0

**Actinides

89 Ac Actinium [227]	90 Th Thorium 232.0	91 Pa Protactinium 231.0	92 U Uranium 238.0	93 Np Neptunium [237]	94 Pu Plutonium [244]	95 Am Americium [243]	96 Cm Curium [247]	97 Bk Berkelium [247]	98 Cf Californium [251]	99 Es Einsteinium [252]	100 Fm Fermium [257]	101 Md Mendelevium [258]	102 No Nobelium [259]

Alkali Metals

1	H	Hydrogen	1.008
3	Li	Lithium	6.941
11	Na	Sodium	22.99
19	K	Potassium	39.10
37	Rb	Rubidium	85.47
55	Cs	Cesium	132.9
87	Fr	Francium	[223]

Alkaline Earth Metals

4	**Be**	Beryllium	9.012
12	**Mg**	Magnesium	24.31
20	**Ca**	Calcium	40.08
38	**Sr**	Strontium	87.62
56	**Ba**	Barium	137.3
88	**Ra**	Radium	[226]

Tranistion Metals

21 **Sc** Scandium 44.96	22 **Ti** Titanium 47.87	23 **V** Vanadium 50.94	24 **Cr** Chromium 52.00	25 **Mn** Manganese 54.94	26 **Fe** Iron 55.85	27 **Co** Cobalt 58.93	28 **Ni** Nickel 58.69	29 **Cu** Copper 63.55	30 **Zn** Zinc 65.39
39 **Y** Yttrium 88.91	40 **Zr** Zirconium 91.22	41 **Nb** Niobium 92.91	42 **Mo** Molybdenum 95.94	43 **Tc** Technetium 98.91	44 **Ru** Ruthenium 101.1	45 **Rh** Rhodium 102.9	46 **Pd** Palladium 106.4	47 **Ag** Silver 107.9	48 **Cd** Cadmium 112.4
* 71 **Lu** Lutetium 175.0	72 **Hf** Hafnium 178.5	73 **Ta** Tantalum 181.0	74 **W** Tungsten 183.9	75 **Re** Rhenium 186.2	76 **Os** Osmium 190.2	77 **Ir** Iridium 192.2	78 **Pt** Platinum 195.1	79 **Au** Gold 197.0	80 **Hg** Mercury 200.6

Boron Elements

5	**B**	Boron	10.81
13	**Al**	Aluminum	26.98
31	**Ga**	Gallium	69.72
49	**In**	Indium	114.8
81	**Tl**	Thallium	204.4
113	**Nh**	Nihonium	[286]

Carbon Elements

6	
C	
Carbon	
12.01	
14	
Si	
Silicon	
28.09	
32	
Ge	
Germanium	
72.61	
50	
Sn	
Tin	
118.71	
82	
Pb	
Lead	
207.2	
114	
Fl	
Flerovium	

Nitrogen Elements

7	N	Nitrogen	14.01
15	P	Phosphorus	30.97
33	As	Arsenic	74.92
51	Sb	Antimony	121.8
83	Bi	Bismuth	209.0
115	Mc	Moscovium	[289]

Oxygen Elements

8	O	Oxygen	16.00
16	S	Sulfur	32.07
34	Se	Selenium	78.96
52	Te	Tellurium	127.6
84	Po	Polonium	[209]
116	Lv	Livermorium	[293]

Halogens

9 **F** Fluorine 19.00
17 **Cl** Chlorine 35.45
35 **Br** Bromine 79.90
53 **I** Iodine 126.9
85 **At** Astatine [210]
117 **Ts** Tennessine [294]

Noble Gases

2 **He** Helium 4.003
10 **Ne** Neon 20.18
18 **Ar** Argon 39.95
36 **Kr** Krypton 83.80
54 **Xe** Xenon 131.3
86 **Rn** Radon [222]

Lathanides

57	58	59	60	61	62	63	64	65	66	67	68	69	70
La	**Ce**	**Pr**	**Nd**	**Pm**	**Sm**	**Eu**	**Gd**	**Tb**	**Dy**	**Ho**	**Er**	**Tm**	**Yb**
Lanthanum	Cerium	Praseodymium	Neodymium	Promethium	Samarium	Europium	Gadolinium	Terbium	Dysprosium	Holmium	Erbium	Thulium	Ytterbium
138.9	140.1	140.9	144.2	[145]	150.4	152.0	157.3	158.9	162.5	164.9	167.3	168.9	173.0

Actinides

89	90	91	92	93	94	95	96	97	98	99	100	101	102
Ac	Th	Pa	U	Np	Pu	Am	Cm	Bk	Cf	Es	Fm	Md	No
Actinium	Thorium	Protactinium	Uranium	Neptunium	Plutonium	Americium	Curium	Berkelium	Californium	Einsteinium	Fermium	Mendelevium	Nobelium
[227]	232.0	231.0	238.0	[237]	[244]	[243]	[247]	[247]	[251]	[252]	[257]	[258]	[259]

SCIENCE CHUNKS

Curie Unit

Marie Curie - Her Life

Marie Curie - Her Work

88

Ra

Radium

[226]

Curie

The Discoverer of Polonium and Radium

1860 1880 1900 1920 1940

SCIENCE CHUNKS

Force Unit

Force

Balance

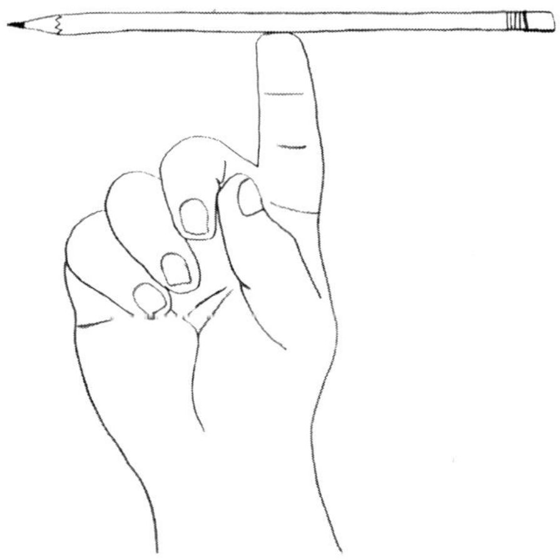

Gravity

Friction

Floating

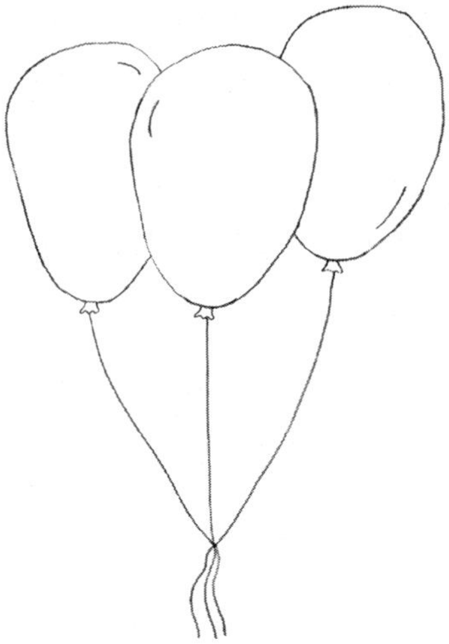

SCIENCE CHUNKS

Motion Unit

Dynamics

Speed

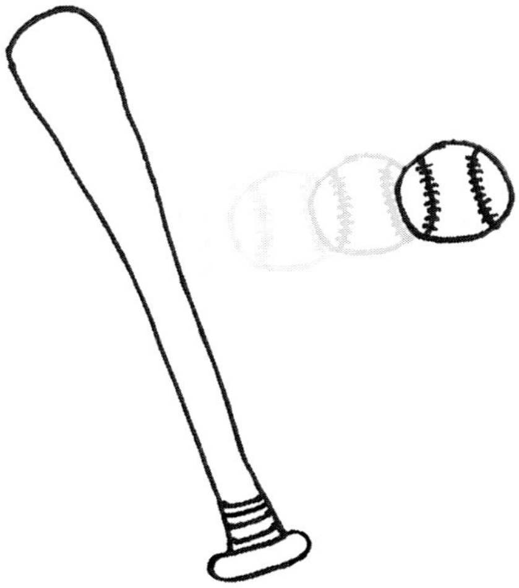

Circular Motion

SCIENCE CHUNKS

Glossary

Year C Glossary

A

Acceleration -

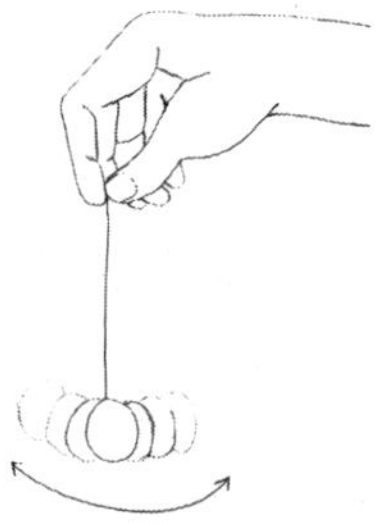

Alloy —

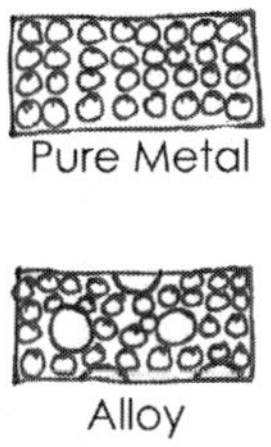

Arthropod —

B

Bacteria —

Balance -

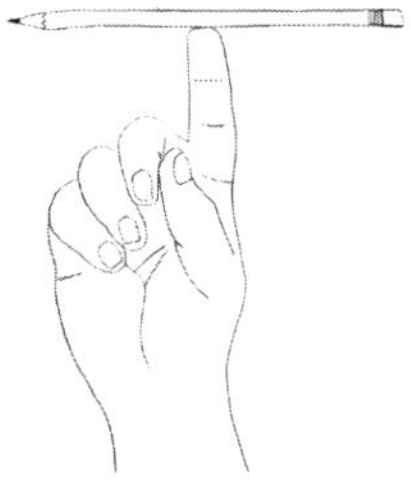

C

Centripetal Force -

Continent —

D

Density -

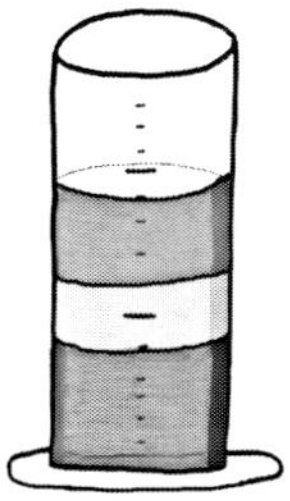

E

Element —

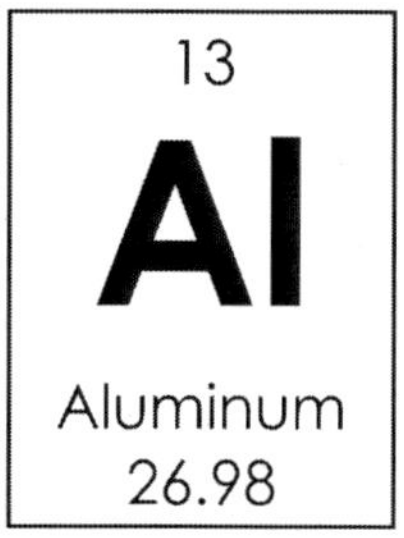

Equator —

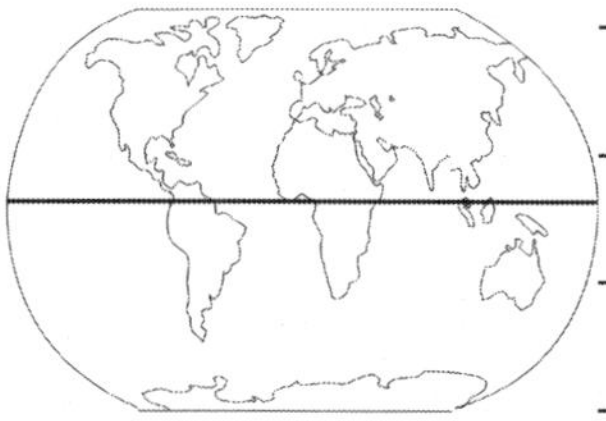

Essential Element —

F

Force -

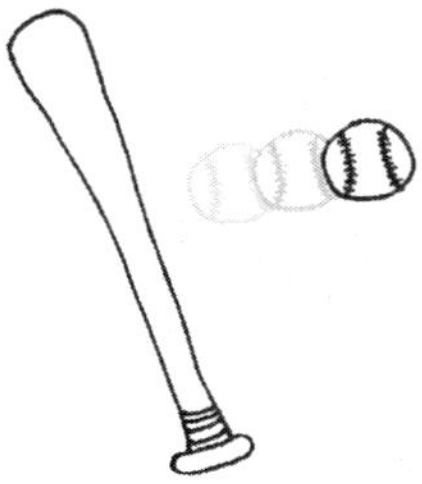

Friction -

Fungus -

G

Gravity -

H

Habitat -

I

Inert —

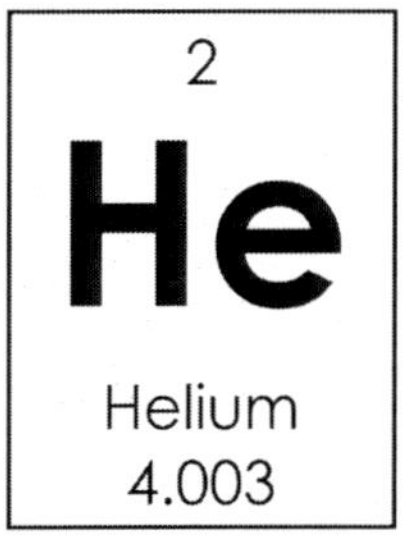

Inertia -

Insect —

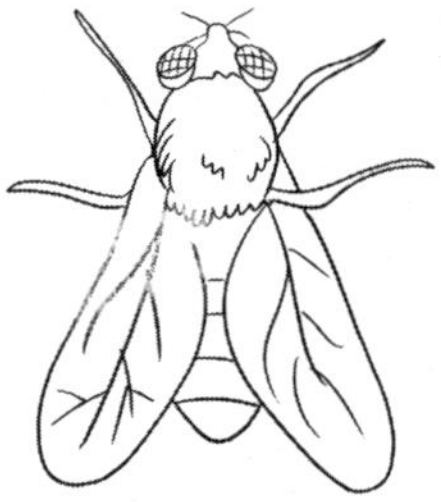

Invertebrate —

Ion —

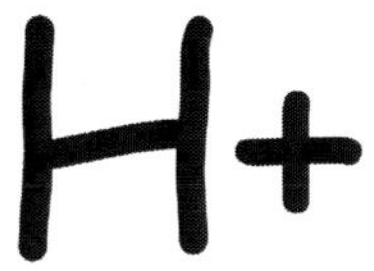

J

K

L

Living -

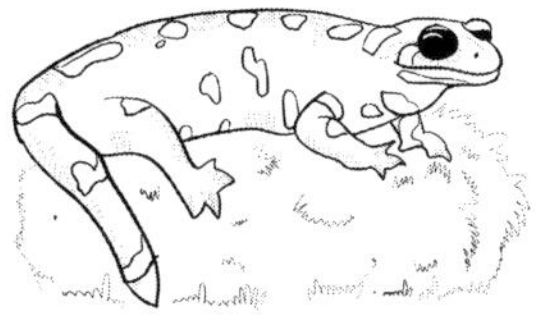

Lubricant -

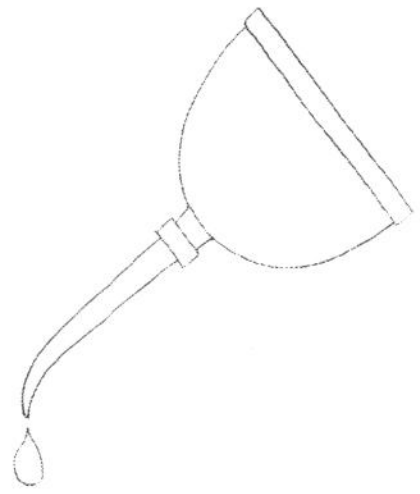

M

Magma —

Magnet —

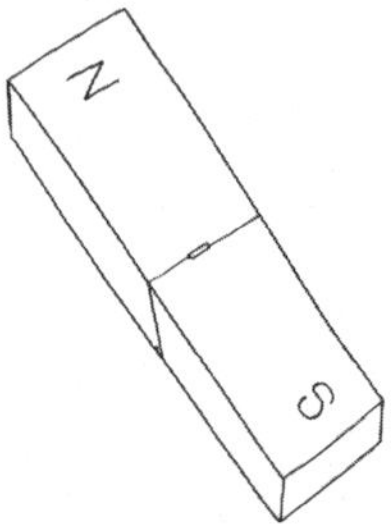

Magnetism —

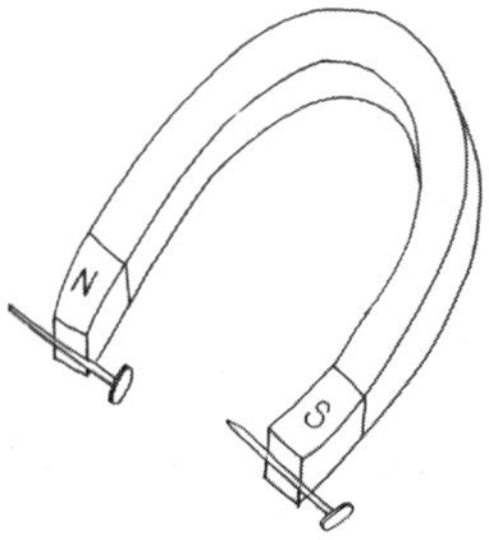

Mass -

Metal —

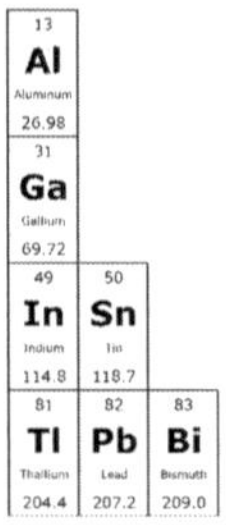

Metalloid —

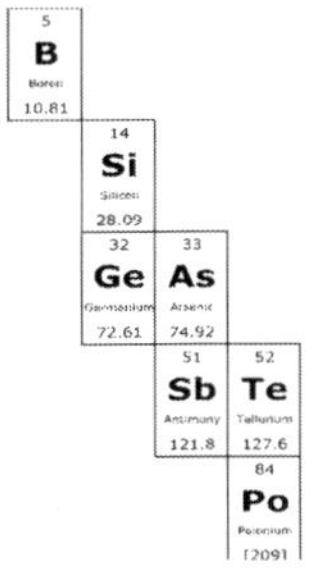

Metamorphosis —

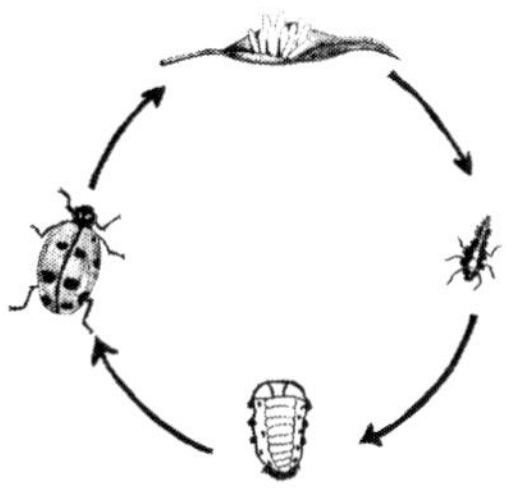

Mollusk —

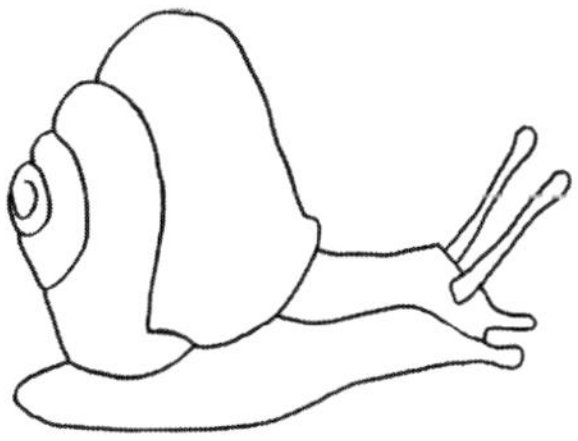

Momentum -

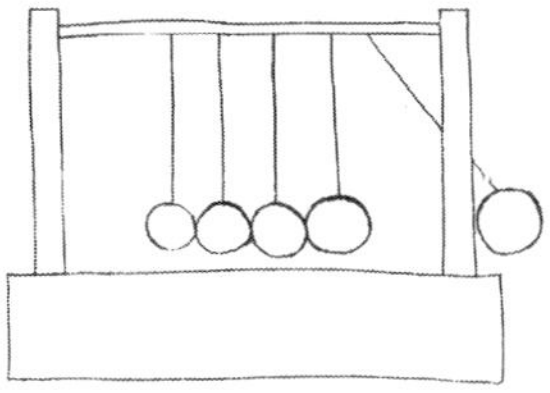

Mountain Range—

N

Nonliving -

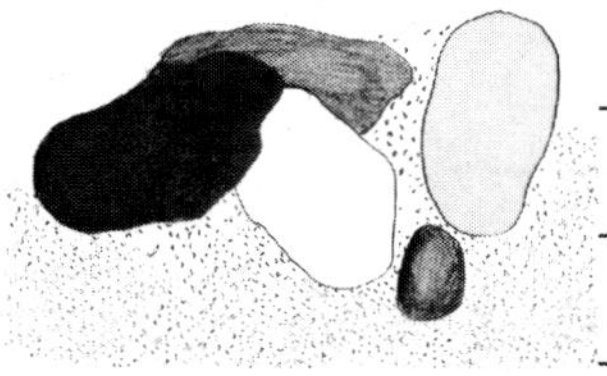

Nonmetal —

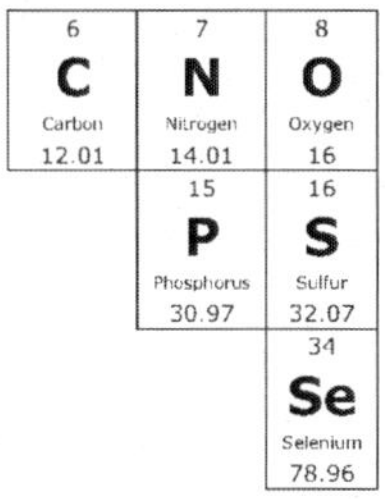

O

Oxidation —

Iron + Damp Air

⇩

Rust

P

Periodic Table —

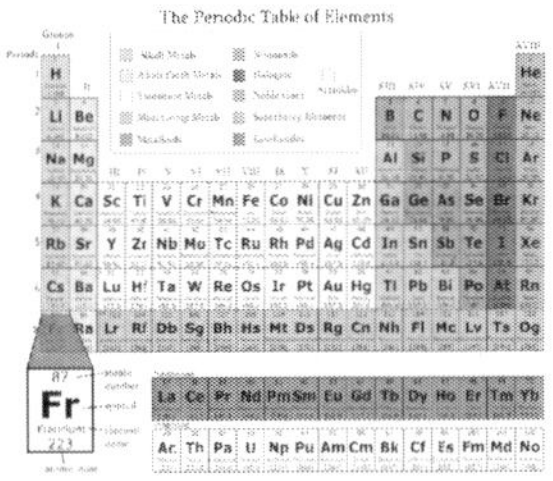

Protist -

Q

R

Radioactive Decay —

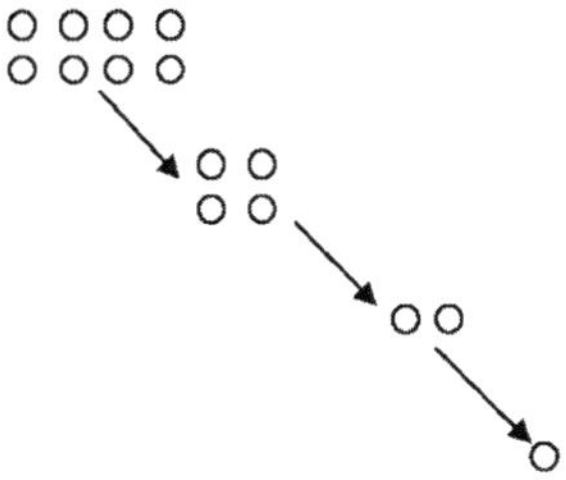

Radioactivity —

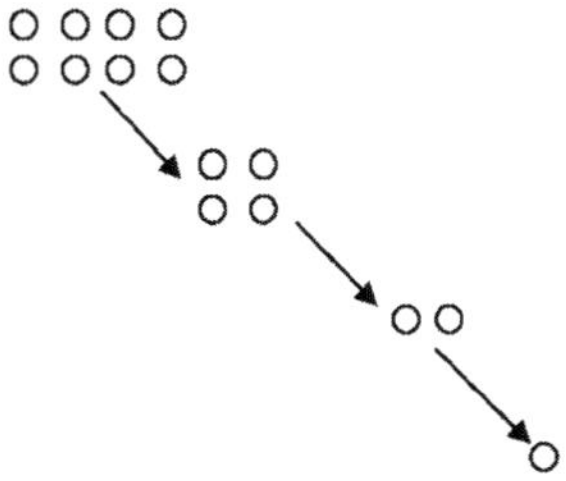

Reactive —

Recycling —

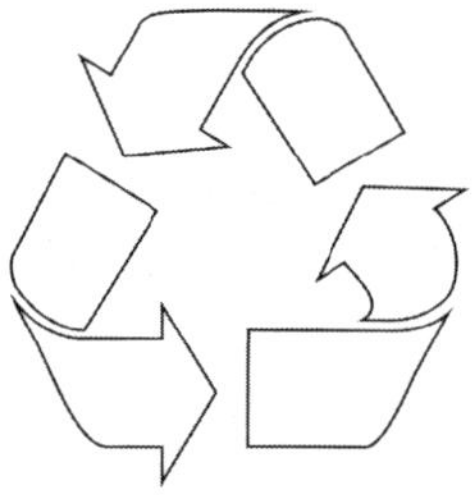

Refraction —

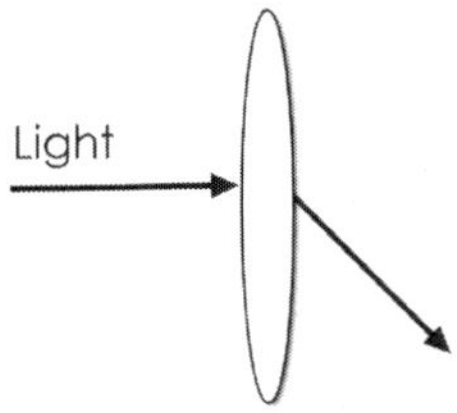

Resultant Force -

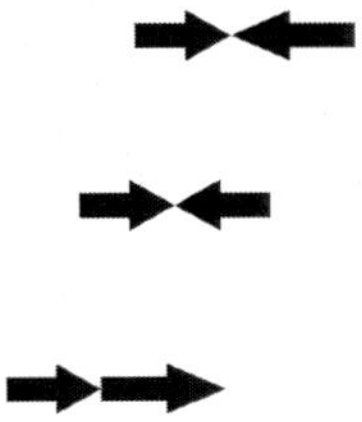

S

Shellfish —

Solution —

Speed -

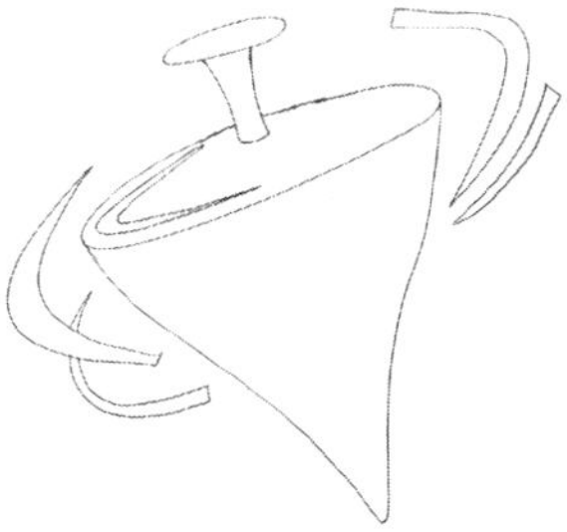

Spore -

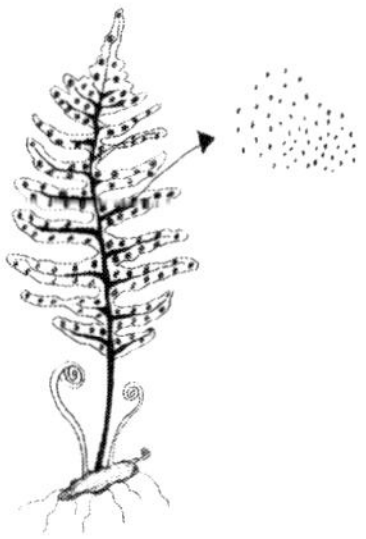

T

Torque -

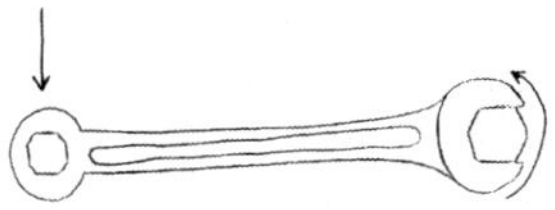

U

V

W

Weight -

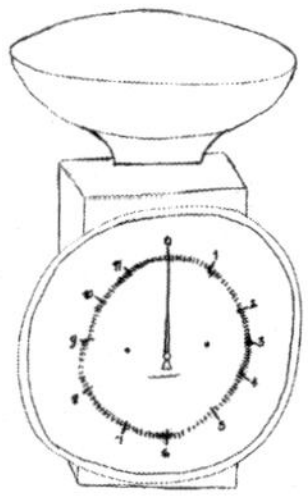

X

Y

Z

Made in United States
North Haven, CT
10 September 2026

18331461R00037